"Kentucky is a place of wonder, a realm of dreams where imagination runs free like wild horses in fields of deep bluegrass."

Kentucky is a place where labor has dignity, music is handmade, family is cherished and people are the greatest resource.

Kentucky is a peaceful land where children play on gentle hills and neighbors share favorite dishes and look forward to harvest time – a time when school resumes and basketball season comes around.

Kentucky is a place where people gather at homecomings and family reunions, at ball games and band camps, at all-day meetings and dinners on the grounds near white churches with tall steeples that paint the landscape with enduring beauty.

Kentucky is a place of timeless farms and noble farmers where season after season, they labor together growing food, families and faith in summer sunshine.

And, somewhere in the middle of it all, lives a very smart little boy with a dream so big and wonderful, if you knew about it, even you would want to see it come true.

This is his story.

"Today is a big day - a wonderful day indeed."

Billy Alexander needed no one to wake him up, even though he never slept a wink the night before. *"Today is a big day,"* he said under his breath, *"a wonderful day indeed."* It was the first day of summer break for young Billy Alexander and today he was going to his favorite place in the whole wide world. *"Today, if Mom and Dad will just wake up, I'm going to spend the summer at Granddad's Kentucky farm,"* he said to his dog, Banjo.

Even though it was still dark outside, Billy was already dressed, sitting on the edge of his made-up bed and staring at the clock on his cell phone. At last, he heard the high-pitched singing voice of his mother, *"Biiiiilleee... time to get up!"*

Like a streak of lightning, he ran for the door, grabbed his bulging suitcase, put on his favorite cap, the blue one that said, *"Young Farmer,"* and ran down the stairs into the bright kitchen. Banjo was right on his heels and crashed helplessly into the back door before he could get stopped.

"*Wow!*" Billy's mother exclaimed, "*Looks like you two are ready to go!*"

"Oh yes we are!" Billy said breathlessly. "*I cannot wait to see Addy! Banjo and I are going to have the best summer ever.*" She poured cold milk over a heaping bowl of cornflakes sprinkled with strawberries and walnuts.

"*I like milk and guess what mom,*" Billy said as white milk splashed on the crisp cereal flakes. "*There are farmers right here in Kentucky with herds that give, on average, 29,000 pounds of milk per cow every single year! Compared to 1950 when they were getting less than 5,000 pounds of milk per cow, that is awesome!*"

"*Billy Alexander, you make my head spin!*" his mother said shaking her head and laughing. Billy was like the Wikipedia of farm facts. He loved school and reading but he especially liked surprising his mom with his latest discoveries about farming. He scooted into his seat and plunged the spoon into the bowl of cereal while Banjo sat right at his elbow, tongue hanging out, hoping for a treat.

"*Don't forget to give thanks!*" his mother said just in time.

Billy let go of the spoon, folded his hands and began to pray,
"*Dear Lord, thank you for Mom and Dad, Grandma and Granddad, Addy and Banjo and summer break, and God, thank you for farmers and food. Amen!*"

Billy always prayed for farmers.

With that, he shoved an overloaded spoonful into his wide, grinning mouth while his mom slid two slices of wheat toast, covered in strawberry jelly in his direction.
"Drink your apple juice," his mom said.
Billy grabbed the glass of juice and poured it down in a single gulp. *"Hey mom, do you know how many apples it takes to make a cup of apple juice? Four!"* Billy asked without giving her a chance to answer. *"And mom, guess how many gallons of water it takes to grow an apple? You won't even believe it!"* he said shaking his head.
"It takes seventeen and one half gallons of water to make one - one single apple! Wow! That means it took about 70 gallons of water to make my glass of juice!"

"Fortunately, here in Kentucky, we have wonderful apple orchards and plenty of fresh water!" His mom replied.

"Mile by mile, the city disappeared behind them"

The neighborhood was still sleeping when Billy's father loaded everything into the car. When the last car door closed, Billy's dad asked, *"Everybody buckled in?"*

"Everybody except Banjo," Billy said laughing. The car pulled carefully onto the busy street and at last, summer break was underway. Mile by mile, the city disappeared behind them. Ancient limestone fences leaned this way and that along the narrow blacktop road while just beyond, beautiful horses stared at passing cars.

"Crossing the river was like going to another world -
the world he loved most of all."

Fog filled the Kentucky River valley as they waited their turn to board the quaint old ferry at Valley View. For Billy, crossing the river was like going to another world - the world he loved most of all.

"All of Billy's favorite memories were at the farm."

Billy had spent every summer at the farm since he was a baby.
All of his favorite memories were there. As the car rolled down the highway, Banjo rearranged himself so he could lay his head on Billy's knee. While the radio played in the background, Billy recalled some of his favorite things about the farm. He remembered the smell of fresh cut hay, the soft chicks he often held in his hands, the bumpy rides in Granddad's old pickup and the hours spent daydreaming with his best friend, Addison Stone, who lived on the farm next to his Granddad.
Billy smiled. After a while Billy spoke up. *"Hey Dad, Friday at school, my friend, Carl, said that he and his sisters would not have enough to eat this week now that school is out. That's awful."*

"That's terrible," his dad expressed agreement.

"It is terrible but not surprising to me, Billy." His mother responded.
"I recently heard that 1 in every 5 Kentucky children don't have enough food. When I get back to the city later today I'll call your teacher and see if I can do something to help Carl and his sisters."

"Thanks mom," Billy said quietly. *"So, I was reading last night and saw an article that said we lost 22 million acres of farmland in the U.S. between 1997 and 2007. That's kind of scary, don't you think? Is that the reason why families like Carl's are hungry?"*

"That may be a part of it," she answered, *"but, there are bigger issues. Still, you need to talk to your Granddad. He tells me that new technologies, disease and insect-resistant crops, and all kinds of improvements in farming – things like genetics and nutrition and science – are enabling fewer farmers to accomplish more and more on less acres. He seems very hopeful."*

Just then, the car began to slow. It turned onto the gravel lane that led to the simple farmhouse. Up ahead, Billy's Grandma was leading a cow with twins up the lane to a pasture closer to the house. In the distance, Billy saw his Granddad, perched high upon the tractor, waving his hat to get Billy's attention.

"Stop!" Billy yelled excitedly.
The car rolled to a stop and Billy dashed away, running and leaping over piles of fresh plowed dirt until he fell breathless into his Granddad's open arms.

Granddad leaned back and examined him head to toe.

"My goodness, Billy!" he exclaimed. *"I think you have grown at least four inches taller since Christmas!"*

Billy couldn't stop smiling as they walked to the house. Then, at that very moment, he heard his best friend, Addison, screaming, *"Billy! Billy!"*

And there she was, jumping up and down on the front porch.

"There is nothing old-fashioned about farming."

When they reached the house, Granddad motioned to Billy's parents with his big, calloused hands, *"Let's go up on the porch and sit a spell.*
Are y'all hungry?
Did you get this boy anything to eat on your way?
Will you be spending the night?"

Billy's mom interrupted the flow of questions.
"Daddy, we can't stay," she said as she hugged him.
"We're not the ones on summer break. We have to get back to the city. Duty calls."
She was saying goodbye to everyone as Billy took his suitcase in the house.
When he got back, the car was already moving away in a cloud of dust as it rolled slowly back toward the city.

Billy and Addy were grinning at each other when
Grandma noticed.
"Are you two scheming already?"
"We want to take a ride on Ol' Joe,"
they said in unison.
"Please?"
"Please?"
"Please?"

"Oh, no!" Grandma answered.
"That horse is so big and so tall and…"

"You could have added, so old and so slow
to your description," Granddad chuckled.
"You don't need to worry about that horse, Grandma.
Ol' Joe is as gentle as he is big.
Besides, I told Billy next time he came
he could ride him and well,
it's time for me to keep my word.
I'll go with them and they'll be fine."

"It's time for me to keep my word."

His bright eyes twinkled as he winked at Grandma and said,
"Go ahead Billy, call him. He'll come for you."

Billy cupped his hands around his mouth like he had seen Granddad do and called out, *"Oooowee! Oooowee! Come Joe!! Come!"*
In the distance, the old horse lifted his massive head from grazing and began moving lazily in the direction of the excited pair.

"Look Billy, you did it!" Addy giggled. *"He's coming! He's coming!"*

Grandma disappeared inside and quickly reappeared with a canvas bag in her hand.
"Here Billy," she said. *"I've got some snacks."*
"Thanks Grandma!" Billy said, pulling the wide strap of the bag over his shoulder.
"Come on, Addy."

Together, they leaped from the porch and ran toward the barn as Banjo ran with them. While they were waiting on the horse, Granddad caught up and soon, Ol' Joe got there, too. Without instruction, he came in close to an old wooden loading chute by the barn, stopped and stood perfectly still. Granddad attached a soft rope to the halter and gave Billy instructions to climb from the loading chute onto Joe's back. When Billy was settled into place, he motioned for Addison to climb on behind him. Joe never moved.

"Okay Addy," Billy said. *"Put your hands around my waist and hold on tight."*
Holding onto the bridle, Granddad clicked his tongue and said,
"Come up, Joe!"
Joe lunged gently forward.

mitchell d. tolle

"Yes indeed! That is the smell of life!"

Granddad and Ol' Joe moved steadily along the mostly grassy lane.
They continued on until they came to a stop on a high ridge overlooking everything.
The farm spread out before them like a patchwork quilt of yellows and greens and lavenders.

The cloudless sky was soft and blue.
In the valley below they could see tiny houses and barns,
fields of half-grown corn and fields freshly plowed, awaiting their seed.
Newborn calves dotted the deep green meadows where
fat cows gorged themselves on tall grass.

Treetops swayed in the gentle breeze.
Billy breathed in deep and filled his lungs with fresh air.
"Do you smell that, Addy?" he asked.
"Yes indeed! That is the smell of life!"

"Did you know that U.S farmers raise nearly 20% of the world's beef with only 9% of the cattle? Farmers like Granddad are growing more and more livestock on less land than ever before in history, and get this, 97% of farms are family owned, just like this one!"
He took another deep breath and said,
"Yes indeed! This is how life smells!"

"It smells a lot like cow chips to me," Addy said giggling.

Granddad chuckled and said, *"Here, let me help you kids down. This is a perfect spot for us to give Joe a rest and see what your Grandma put in that bag!"*

The cows stared curiously as they slid down from the big horse and sat cross-legged together in soft grass as Granddad sorted through the big canvas bag. There were apples and pears, jelly biscuits left over from breakfast and several bottles of spring water.

"You kids take whatever you want," Granddad said as he took a big bite out of one of the biscuits. *"This one's for me!"*

Ol' Joe had moved a few feet away to where the grass was deeper and started grazing again. Billy took apples from the bag, gave one to Addy and then took a bite of his own. *"I love coming here,"* Billy said to no one in particular. *"I can't imagine what it would be like to live on a farm all the time."*

"Well I can!" Addy said smiling. *"Three words describe life on a farm. They are: 'Fun, fun and more fun!' "* Anytime Addy spoke, you could hear a smile in her voice.

"I know they say farm work is hard, and you say it's noble and important, and mom calls what I do 'chores', but, Billy, I don't feel like they are chores at all. I like feeding chickens and gathering eggs. I like getting fresh food from the garden and helping mom can green beans, pickled beets, tomatoes and sweet corn. I really, really like going with dad to steal honey from the beehive and I like climbing the apple tree to pick my own special apple. To me, farming is an endless adventure!" She laughed out loud.

"Let's just face it Mr. Alexander," her face shined with joy, *"I am a country girl and country girls are bound to be farmers!"*

"Farming is high-tech, cutting edge and super amazing."

As he listened to the kids, Granddad looked out over his farm and smiled.

"Addy, just hearing about how much you like farming makes me want to be a farmer even more," Billy said wistfully.

"Really? Addy said. *"I thought you might think I was kinda' old-fashioned."*

"No, I don't." Billy replied seriously.
"You are going to be a great farmer. Besides, there is nothing old-fashioned about farming. The advancement in farming is mind-boggling. A farmer today can grow 120% more soybeans, 230% more corn and 200% more wheat compared to a farmer in 1960! And, listen Addy, the same thing is true for chickens."

"A hen today can lay about 276 eggs each year, which is 60% more than in 1950 and, broiler chickens now produce 81% more meat! Farmers are learning more and getting better every day! Addy, it is not old-fashioned. Farming now is high-tech, cutting-edge and super amazing!"

Addy stood and put her hands on her hips and said,
"Well, Mr. Alexander, it looks like this girl is gonna be a Supa' Farmer!"
Banjo barked and Billy clapped his hands in approval.

"It does have a nice ring to it," Billy said.
"Addison Stone: Super Farmer!"
They laughed and laughed.

And while they laughed, Banjo chewed on a stick.

Inspired by what he was hearing, Granddad turned toward them and said, *"A lot more grows on a farm than chickens, cows and corn. And just listening to you two reminds me just how true that is on this farm. Farming is about more than making a living, too. Farming is a way of life. Here on the farm we grow food but we also grow families. The things for which I'm most thankful are not out there somewhere in a field but right here next to me. I'm so proud of you kids."*

As they had talked, unnoticed by the kids, rain clouds had drifted in from the west and filled the once cloudless sky. Granddad said, *"Hey kids, we better get back to the house before the rain gets here."* The kids jumped to their feet and quickly gathered everything back into the bag. The wind stirred more intensely, signalling that things were changing rapidly. Granddad helped each of them up onto Ol' Joe and turned him toward home. His steps were a little quicker as they scrambled to beat the rain. Distant thunder and the hurried pace caused Addy to squeeze Billy's waist a little tighter than before.

By the time they had set Ol' Joe free, large drops of rain were falling like marbles into the dust. Banjo danced in wonder around the kids as they lifted their arms high and opened their mouths wide to let the rain fall on their tongues and their faces.

As the rain fell even harder, they ran in circles, in no real hurry to get to the house. Just as they leaped upon the porch, the heavens opened and it began to pour, making a deafening sound on the porch's metal roof.

A few steps behind, Granddad rushed up the steps, out of breath and dripping wet. He laughed at the scene and turned away just in time as Banjo shook himself, giving everyone an unwanted shower.

Hearing all the clatter, Grandma came through the screened door with an armload of towels. *"Well, aren't you a pretty bunch,"* she said giving each one a towel.

As she turned to go back inside, Billy heard her say faintly, *"Thank you God for the rain. We sure needed it."*

Once inside, Granddad went straight to his rocker, sat down and began to take off his work boots. Billy and Addy went to the other side of the living room where they sank down deep into the soft, plush couch.

Grandma headed off into the kitchen to begin preparing dinner. Granddad leaned back in his rocker and was soon lulled to sleep by the rhythm of the rain on the old tin roof.

"Addy, I'm not sure if you know it, but I'm going to be a farmer, too."

"Wow! That's so cool Billy. You and me – Supa' Farmers! We'll wear masks and capes and save the world just like real supa' heroes!" The room filled with her laughter.

"Well," Billy said with his most serious sounding voice. *"My reason for becoming a farmer is not because I'm looking for adventure and it's not because I want to have fun, even though farming can be a lot of fun. Addy, I actually know kids who don't have enough food to eat. I've heard about people in foreign lands struggling to get food, but this is right here, in my school and it's my friend, Carl, and his sisters."* You could hear the concern in his voice.

Billy twisted in his seat and continued without noticing that his Grandma was now standing in the doorway taking it all in.

"Addy, the need is way too big for two farmers – even super farmers." You could feel his intensity. *"This will require all of us working together. My dream is to plant more than corn and soybeans. I want to plant farms and young farmers all over Kentucky. We will think of farms as seeds, growing and multiplying in every valley and on every hillside all over the Bluegrass land.*

If you will accept it, Addison Stone, this is our mission!"

"To me, farming is an endless adventure."

"I accept Billy Alexander! I accept!" Addison was ecstatic.

"Listen Addy, the average farmer in Kentucky is 58 years old!" Addy's mouth fell open in disbelief.

"Really? Do our parents know that?" she asked.

"And get this," Billy continued. *"About 10,000 people become 65 years old every day in America! Addy, this is a crisis! Who will take the place of farmers like Granddad?"*

Addy jumped to her feet. *"We will! We will! We'll do it! We will become farmers like Granddad and recruit young farmers and plant farm seeds everywhere and feed kids like Carl and save the world!"*

"Oh! Billy, this is so exciting," Addy's mind was racing. *"Billy Alexander, this is the most wonderful dream I've ever heard. When do we start? What do we do first? Oh! Billy, dreams change everything and a dream like this can change the world - for good!"*

While Billy and Addy were pondering their dream, Grandma said, *"Hey everybody, thanks to a farmer - it's time to eat!"*